FINDING DINOSAURS

SPINOSAURUS

by Samantha S. Bell

FOCUS READERS

FOCUS READERS

WWW.FOCUSREADERS.COM

Focus Readers is distributed by North Star Editions:
sales@northstareditions.com | 888-417-0195

Produced for Focus Readers by Red Line Editorial.

Content Consultant: Dr. David B. Weishampel, Professor Emeritus, Center for Functional Anatomy and Evolution, Johns Hopkins University School of Medicine

Photographs ©: Kostyantyn Ivanyshen/Shutterstock Images, cover, 1; Jesse33/Shutterstock Images, 4–5; Red Line Editorial, 7; Piero Oliosi/Polaris/Newscom, 8, 26–27; Francois Mori/AP Images, 10–11; Ingo Schulz/ImageBroker RM/Glow Images, 13; Arthur Dorety/Stocktrek Images/Alamy, 15; Kay Nietfeld/picture-alliance/dpa/AP Images, 16, 29; Bao Dandan/Xinhua/Alamy Live News, 19; LindaMarieB/iStockphoto, 20–21; Jake Kohlberg/Shutterstock Images, 23; ineersk/Shutterstock Images, 25

ISBN
978-1-63517-505-9 (hardcover)
978-1-63517-577-6 (paperback)
978-1-63517-721-3 (ebook pdf)
978-1-63517-649-0 (hosted ebook)

Library of Congress Control Number: 2017948065

Printed in the United States of America
Mankato, MN
November, 2017

ABOUT THE AUTHOR

Samantha S. Bell is an author and illustrator from South Carolina who has written more than 60 nonfiction books for children. She loves learning about dinosaurs and how they may have lived. She also enjoys being outside and digging in the dirt, but she has yet to discover any dinosaur bones.

TABLE OF CONTENTS

CHAPTER 1

BONES IN A BOX

In 2008, Nizar Ibrahim was searching for dinosaur fossils in Morocco. He visited rock formations known as the Kem Kem Beds. In a town called Erfoud, Ibrahim purchased a box of fossils from a local fossil hunter. The box held several bones stuck inside blocks of purplish stone.

Many people travel to Erfoud to purchase fossils from local fossil hunters.

The next year, Ibrahim went to a museum in Italy. Researchers there showed him a partial skeleton of a large dinosaur they had recently bought. The skeleton included leg bones, ribs, **vertebrae**, and several long spines. Ibrahim was astonished. It was clear from the bones' shape that they came from a Spinosaurus.

This was an important discovery. The only other Spinosaurus specimen had been found in Egypt by Ernst Stromer in 1911. The bones had been placed in a museum in Germany. But during World War II (1939–1945), the museum and the fossils were destroyed. Scientists only knew about Spinosaurus bones from Stromer's notes and drawings.

Ibrahim recognized the rock around the fossils, too. His box from Morocco contained the same purplish stone.

MAP OF SPINOSAURUS FINDS

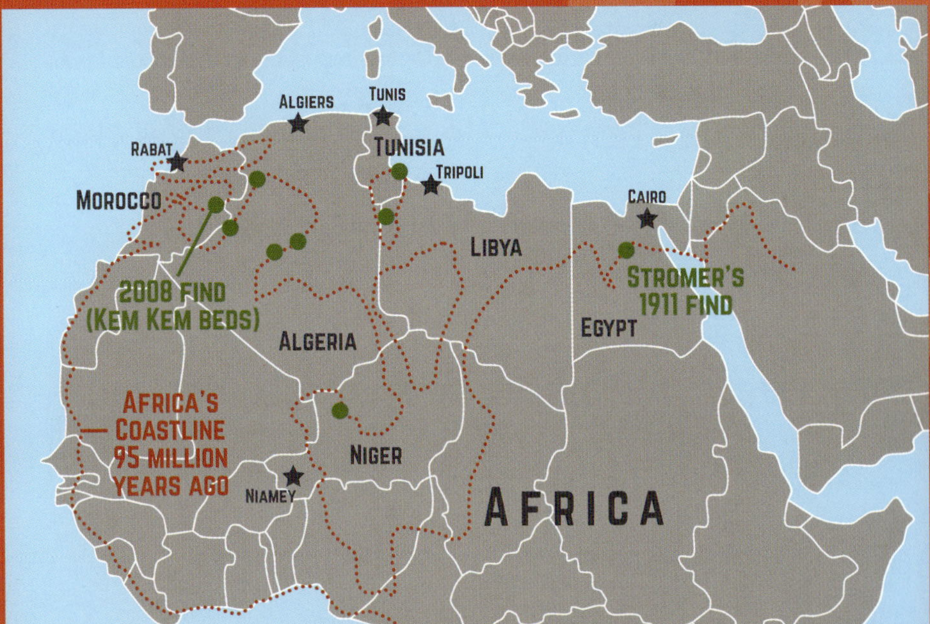

Nizar Ibrahim (left) helped uncover the Spinosaurus fossils.

Ibrahim realized that his fossils were also Spinosaurus bones. He wanted to find the fossil hunter and ask if the two sets of fossils came from the same site. Maybe they were even part of the same dinosaur.

In 2013, Ibrahim returned to Erfoud and found the fossil hunter. The man confirmed what Ibrahim had guessed. The bones in the box and the bones in

Italy came from the same hillside. He agreed to show Ibrahim this site. There, Ibrahim found more Spinosaurus fossils, including parts of a skull and some vertebrae. Together, these fossils would help scientists learn much more about Spinosaurus.

THE KEM KEM BEDS

The Kem Kem Beds are a rock formation located in southeastern Morocco. They cover a part of the Sahara Desert that is 150 miles (241 km) long. It is an area of sandstorms, snakes, and scorpions. Fossils of many prehistoric predators have been found in the Kem Kem Beds' reddish rock. Most fossils are small fragments. But local fossil hunters pick up teeth and bigger pieces to sell to shops.

CHAPTER 2

BUILT FOR LAND OR WATER?

The Spinosaurus skeleton in Italy was much more complete than Stromer's lost specimen. Combined with Ibrahim's bones, it formed a more complete Spinosaurus skeleton than anyone had seen before. Scientists now knew more about what the dinosaur looked like.

The name *Spinosaurus* means "spine lizard."

Spinosaurus stretched 50 feet (15 m) from its nose to its tail. That is approximately 10 feet (3.0 m) longer than the largest Tyrannosaurus ever found. Spinosaurus may even be the largest of all the **carnivorous** dinosaurs.

Spinosaurus had short back legs and small hip bones. It also had a long body and a long neck. Because of these unusual **proportions**, Spinosaurus probably could not walk on two legs like other carnivores. It would have fallen forward. Instead, Spinosaurus most likely had to walk on all fours.

Long spines grew along the dinosaur's back. The spines were probably joined

Spinosaurus's spines could be 7 feet (2.1 m) tall.

together by skin to form a huge sail. Scientists are not sure how this sail was used. It might have helped the dinosaur appear larger and **intimidate** enemies. It could have been used to attract a mate.

A few scientists think the sail might have helped the dinosaur control its body temperature.

Some scientists believe Spinosaurus was more suited for water than for land. It could have used its short back legs to paddle through the water. Other scientists are not so sure. The dinosaur's

SAILS AS SIGNALS?

Even when Spinosaurus hunted underwater, it could still be seen. Its huge sail would rise above the surface, much like a shark fin. The sail might have shown other dinosaurs how big and how old the Spinosaurus was. That way, a smaller Spinosaurus would be able to avoid a larger one.

Spinosaurus might have hunted fish in the water.

center of mass would have made floating with its head above water difficult. It would not have been able to swim in the deep. Instead, Spinosaurus may have lived partly on land and partly in water. That would make it the first **semiaquatic** dinosaur ever found.

The bones in a Spinosaurus foot were long and strong.

Spinosaurus bones seem more similar to the bones of aquatic animals than the bones of other dinosaurs. Spinosaurus bones were very dense. Modern aquatic animals such as manatees and penguins have dense bones, too. Dense bones help these animals float in the water.

There are other clues that Spinosaurus spent time in the water. For instance, the bones in its tail were connected loosely. That means its tail might have moved back and forth like the tail of a fish.

Spinosaurus had long, flat claws on its back feet. Some scientists think the dinosaur had webbed feet, similar to a duck. Webbed feet would have helped Spinosaurus swim through the water or walk on soft mud. Spinosaurus also had powerful front legs with long, curved claws. The claws could hook slippery prey in the water.

FLEXIBLE JAWS

Scientists studied fossils of Spinosaurus jaws. This study revealed an unusual trait. The fossils show ridges where the jaw bones come together in the front. **Tissues** were attached at the ridges. This means the bones were not **fused** together. They could move apart and become wider.

In addition, the lower jawbone was not attached directly to the skull. Instead, it was connected by a flexible joint. This allowed the dinosaur's mouth to open very wide, much like the mouth of a snake.

With its moveable jaw, Spinosaurus could have eaten large prey quickly. It might have opened its mouth and swallowed its prey whole. If so, it is the first time this feeding style has been discovered in dinosaurs. Some scientists think Spinosaurus would have been able to swallow

Spinosaurus probably ate reptiles, fish, and sharks.

turtles and crocodiles. It may have even eaten small dinosaurs.

CHAPTER 3

FEASTING ON FISH

Spinosaurus lived 145.5 to 65.5 million years ago during the Cretaceous period. At this time, the Kem Kem Beds were covered with thick plants. They were part of a tropical river system. The river stretched from Morocco to Egypt.

But Spinosaurus was not the only large carnivore that lived in this area.

As one of the largest predators, Spinosaurus was probably at the top of the food chain.

Scientists have found fossils of sharks, flying reptiles, and crocodile-like creatures. Other predators, including close relatives of Tyrannosaurus, lived there was well. In fact, some scientists think this area may have been one of the most dangerous places on the planet.

THEROPODS

Spinosaurus was part of a group of dinosaurs called theropods. *Theropod* means "beast-footed." The group also includes Tyrannosaurus and Velociraptor. But Spinosaurus was different from the other dinosaurs in this group. Most theropods had hollow, thin-walled bones. They walked on two legs and had sharp teeth for tearing their prey. Spinosaurus had none of these features.

Spinosaurus's teeth have a shape similar to the teeth of animals that eat fish.

Spinosaurus probably ate giant fish and sharks. It used its teeth to grab prey. The teeth of most carnivores are shaped like **serrated** knives. But Spinosaurus's teeth were shaped like cones instead.

Each tooth had only one sharp, smooth edge. This shape was much like the teeth of alligators and crocodiles.

Spinosaurus was similar to a crocodile in other ways. For instance, Spinosaurus had a long, narrow snout. Small nostrils sat in the middle of its skull. These features would have allowed the dinosaur to breathe while part of its head was in the water. Spinosaurus may have **submerged** itself like a crocodile.

Spinosaurus also had openings in the front of its snout. Scientists think these openings held pressure sensors. These sensors would have helped Spinosaurus detect movement underwater. The

Spinosaurus's teeth fit together like a crocodile's teeth do.

dinosaur could have used the sensors to find prey in murky water or mud. Crocodiles and alligators have similar sensors on the ends of their snouts.

CHAPTER 4

MORE TO DISCOVER

Spinosaurus went **extinct** at the end of the Cretaceous period. Many scientists believe a large meteor hit Earth at this time. But no one knows for sure.

Because few Spinosaurus fossils have been found, scientists are not certain what the dinosaur looked like. Ibrahim and a team of scientists tried to find out.

Paleontologists continue to dig for Spinosaurus fossils in Morocco.

They scanned the Spinosaurus bones and studied Stromer's notes and diagrams. Then they used computers to build a digital Spinosaurus skeleton. A 3D printer helped them create a life-size model.

But other researchers thought the model was not accurate. They pointed out that Ibrahim's team used some bones from a teenage Spinosaurus and some

CONFUSING FOSSILS

Many fossils in the Kem Kem Beds were found by local fossil hunters, not **paleontologists**. The fossil hunters do not collect any data with their finds. This makes it difficult for scientists to figure out which species actually lived together.

A Spinosaurus skeleton is on display at the Museum for Natural History in Germany.

bones from an adult. The team also used bones from similar dinosaur species such as Suchomimus to fill in the skeleton's missing parts, including the arms and hands. Even the original bones from the Kem Kem Beds might have come from more than one species. In the future, a new find may reveal the secrets of this strange dinosaur.

FOCUS ON
SPINOSAURUS

Write your answers on a separate piece of paper.

1. Write a paragraph describing how Spinosaurus was similar to a crocodile.
2. If you found a Spinosaurus bone, would you keep it, sell it, or give it to a museum? Why?
3. Where were the first Spinosaurus bones found?
   - **A.** Morocco
   - **B.** Egypt
   - **C.** Italy
4. Why would paleontologists use the hands from a different dinosaur to reconstruct a Spinosaurus skeleton?
   - **A.** They wanted to create a new dinosaur species.
   - **B.** They assumed the hands were similar since the dinosaurs were closely related.
   - **C.** They hoped the hands would make Spinosaurus look more ferocious.

Answer key on page 32.

GLOSSARY

carnivorous
Hunting and eating other animals.

extinct
No longer living on Earth.

fused
Joined together to form one unit.

intimidate
To scare or threaten something or someone.

paleontologists
Scientists who study the ancient past and the fossil remains of ancient living things.

semiaquatic
Living or growing partly on land and partly in water.

serrated
Having a jagged edge, like a saw.

submerged
Covered with water.

tissues
Groups of similar cells in a plant or animal that have a certain job or function.

vertebrae
The small bones that link together to form the backbone.

TO LEARN MORE

BOOKS

Gray, Susan H. *Spinosaurus.* Mankato, MN: The Child's World, 2015.

Hoena, Blake. *Everything Dinosaurs*. Washington, DC: National Geographic, 2014.

Korpella, Bob. *Dinosaur Dictionary for Kids: The Everything Guide for Kids Who Love Dinosaurs*. Waco, TX: Prufrock Press, 2016.

NOTE TO EDUCATORS

Visit **www.focusreaders.com** to find lesson plans, activities, links, and other resources related to this title.

INDEX

Answer Key: 1. Answers will vary; **2.** Answers will vary; **3.** B; **4.** B